A Guide to Chip-Carving Gourds

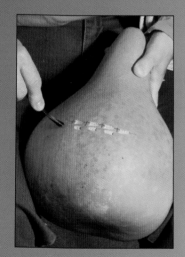 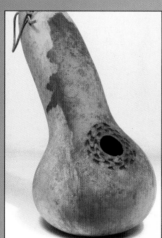  

Marilyn Rehm

4880 Lower Valley Road, Atglen, Pennsylvania 19310

Acknowledgments

I am very fortunate to be in the middle of three generations of family who have worked with gourds; I have been continually encouraged to grow, show, and teach. The American Gourd Society publishes a full color magazine, and state chapters sponsor gourd shows with workshops to introduce gourds to the public. With the recent increased interest in gourds, growers and seedsmen have worked to increase the variety and quality of gourds available to crafters.

A special thanks to David Stichweh for his patience as well as technical skill in photographing the carving projects in this book.

Other Schiffer Books on Related Subjects
Coiled Designs for Gourd Art, 978-0-7643-3011-7, $14.99
Gourd Art Basics: The Complete Guide to Cleaning, Preparation, and Repair, 978-0-7643-2829-9, $14.95
Gourd Crafts: 6 Projects & Patterns, 978-0-7643-2825-1, $14.95
Decorating Gourds: Carving, Burning, Painting, & More, 0-7643-1312-6, $14.95

Designed by Mark David Bowyer
Type set in Freeform721 Blk BT / NewBaskerville BT

ISBN: 978-0-7643-3210-4
Printed in China

Schiffer Books are available at special discounts for bulk purchases for sales promotions or premiums. Special editions, including personalized covers, corporate imprints, and excerpts can be created in large quantities for special needs. For more information contact the publisher:

Published by Schiffer Publishing Ltd.
4880 Lower Valley Road
Atglen, PA 19310
Phone: (610) 593-1777; Fax: (610) 593-2002
E-mail: Info@schifferbooks.com

For the largest selection of fine reference books on this and related subjects, please visit our web site at:
www.schifferbooks.com
We are always looking for people to write books on new and related subjects. If you have an idea for a book please contact us at the above address.

This book may be purchased from the publisher.
Include $5.00 for shipping.
Please try your bookstore first.
You may write for a free catalog.

In Europe, Schiffer books are distributed by
Bushwood Books
6 Marksbury Ave.
Kew Gardens
Surrey TW9 4JF England
Phone: 44 (0) 20 8392-8585; Fax: 44 (0) 20 8392-9876
E-mail: info@bushwoodbooks.co.uk
Website: www.bushwoodbooks.co.uk
Free postage in the U.K., Europe; air mail at cost.

Contents

Introduction

Gourd shapes are so elegant that they really need no enhancement, but if decoration is desired, then carving is ideal since it chisels the natural shell of the gourd into delicate patterns. Carving with hand gouges is a quiet, dust-free process, not constrained by an electric cord. Hand-carving on gourds is typically done with U-shaped wood gouges, and it's a very different process from chip-carving done on wood with a set of knives.

Indonesian bottle gourd

Indonesian bottle gourd with carving

Some really thick-shelled African gourds with the outer layer of shell abraded have been chip-carved in traditional fashion, but the thickness of the shell on most gourds grown in the United States does not permit the depth of carving that is necessary for patterns that rely on the play of light and shadows. Although U-shaped gouges can also be used to produce relief-style carving, chip-carving on gourds is commonly understood to mean geometric patterns of swirls. While applying the term "chip-carving" to gourds that are carved with U-shaped gouges may be a misnomer so far as wood carvers are concerned, the term is widely accepted in the gourd world, and to be sure, it does make "chips" fly!

Relief carving done with U-shaped gouges, background removed

This gourd from Africa was chip-carved with traditional knives.

Relief carving done with U-shaped gouges, tree and bird removed

"Geometric" is a key word here, since the patterns are laid out with a straightedge and compass. Designs will show triangles, rectangles, circles, and combinations of these elements. Carving does not take much force because a gentle rocking motion of the gouge propels it forward. Wood carvers who have switched to gourds say that gourds are easier to carve than wood since only the outer layer of the gourd shell is hard, and most of the thickness of the shell is lightweight fibrous material; also there is no "grain" to the gourd shell. The odd shapes of the gourds may make them harder to hold than a block of wood that can be put in a vise, but since it only takes one hand to carve, the other hand and a lap are sufficient to steady the gourd.

The carving projects in this book are done on gourds with many shapes and sizes, but use only a single ¼" U-shaped gouge and just one simple staining technique. The first project – carving a band along the edge of a bowl– illustrates basic pattern layout, cutting, and staining — and serves as a reference for more advanced projects. The second band project covers birdhouse construction details, and each successive project introduces a new pattern and/or technique. These projects need not be done in order, and the patterns may easily be adapted to other kinds of gourds, producing a variety of bowls, birdhouses, whole carved gourds, and ornaments for a very small investment in equipment.

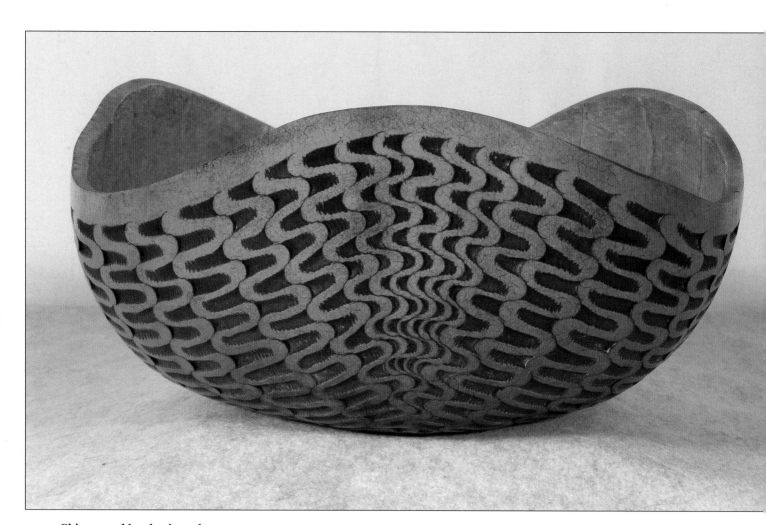

Chip-carved bowl, triangular pattern

**Chip-carved bowl,
rectangular pattern**

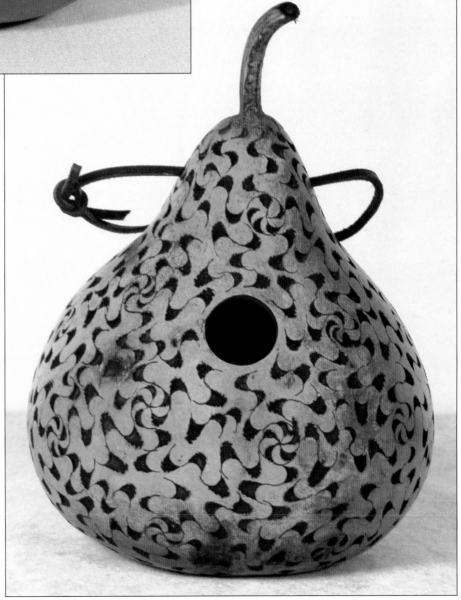

**Chip-carved birdhouse,
circular pattern**

Chip-carved bird

Chapter One:
Getting Started

Selecting Gourds

A sturdy clear-skinned dry gourd is ideal for carving. A good carving gourd will feel heavy for its size and will usually have a nice stem. Test the ability of a gourd to withstand the pressure of the gouge by exerting pressure with your thumbs all over the gourd. There should be no "give" at all as you examine the gourd. Gourds can vary tremendously in thickness; a good carving gourd does not have to be especially thick, but it should be very hard to the touch. Beginning carvers tend to carve more deeply than necessary, so gourds grown in the south or west (with the longer growing season producing thicker shells) are a good choice until a lighter carving touch is developed.

Apply pressure from the thumbs to test the gourd's sturdiness.

Shown is a really thick gourd.

This gourd is thinner, but it's still good for carving.

Since the spaces between the chips actually form the pattern, the clearer the shell, the better the contrast will be between the chips and the background. In the natural process of drying, mold may etch blotches that cannot be removed when the gourd surface is cleaned. A light mold mosaic is not usually a problem, but try to avoid dark mold marks. It's also best to avoid gourds that have a whitish, rough look to the shell, since this indicates that the integrity of the shell has been compromised and these gourds will often stain very dark, obscuring the carved pattern. Bug bite scars or cracks will also stain dark.

Insect damage shown before staining

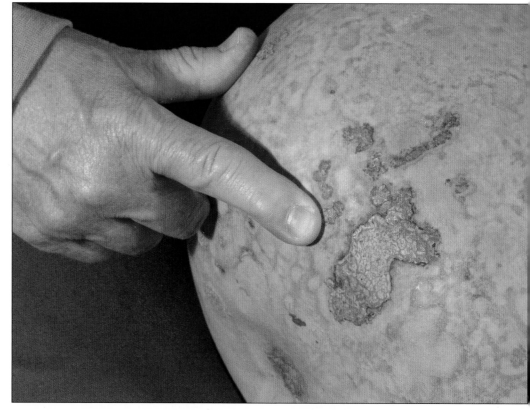

Insect damage stains very dark

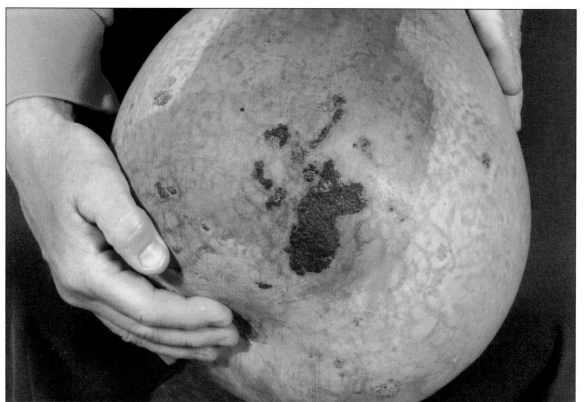

If you buy dry gourds that are already clean, any flaws will be evident, but unclean gourds are much less expensive (especially if homegrown), and often it's possible to work around problem parts of the gourd exposed by cleaning. To remove the outer skin and mold, soak the gourd in a bucket of warm water for 15-20 minutes and then scrape the gourd with the backside of a table knife or a plastic kitchen scrubbie.

Using steel wool or rough sandpaper to clean a gourd will likely result in fine scratches that are not noticeable before staining, but will pick up the stain and leave marks when leather dye is applied. Any outer skin left on the gourd by accident will stain lighter than the background and is really hard to remove after staining without causing a mottled appearance.

Scrape off mold.

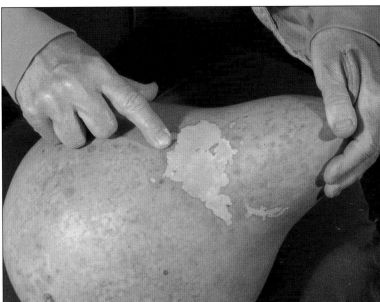

A patch of adhered outer skin before staining

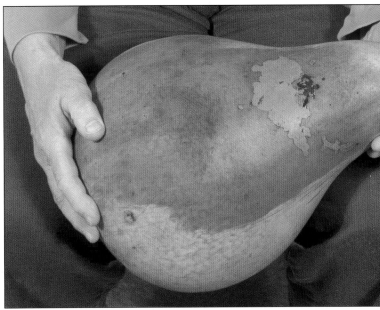

Adhered skin after staining; also note sandpaper scratches to the left of the skin patch.

Selecting Tools

The gourds in this book were carved with a U-shaped wood gouge, which carves a chip about ¼" wide. Not all tool manufacturers measure the width of gouges in exactly the same way. If the ¼" is measured across the tip of the gouge, the chip will be slightly larger than if the measurement is made part way down the tool profile. An ideal way to choose a gouge is to go to a large hardware store or a shop that specializes in wood carving tools and actually handle the tools. In addition to the width of the gouge, a second number associated with a U-shaped gouge is the sweep number that describes the depth of the "U" as you look straight on at the tool's cutting edge. The sweep measurements are not standard across all brands, but generally a half-circle is considered to be a #9 sweep and the next size "taller" than a half circle profile would be a #11 sweep. A shallower, more bowl-shaped tool profile, would be a #5 or a #3 sweep. All of these sweeps can be used, but a tall "U" like the #11 is ideal for a beginning carver. Using a gouge with a sweep other than a #11 will make the patterns shown here look slightly different — not bad, just slightly different.

quiring a longer shaft and handle to keep your hand from bumping into adjacent parts of the gourd.

Prices of gouges can also vary considerably based on the quality of the metal of the cutting edge. A higher-priced gouge should keep a sharp edge longer, but a moderately priced gouge will work fine on gourds if sharpened occasionally by running it back and forth on a sharpening stone.

Many carvers choose to wear a protective glove on the hand that holds the gourd. Special flexible mesh gloves are available at woodworking supply stores, but a work glove with a non-slip surface is quite suitable. If you choose to carve barehanded, keep the band-aids handy! A piece of waffle-weave rubber shelf liner (designed for use in camper-trailers) works well in your lap to keep the gourd from slipping — and also to protect the fabric of your clothing from being pulled and worn as you continuously turn the gourd in your lap.

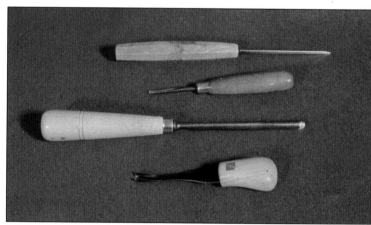

Various handle styles

Tool profile of a U-gouge

Another factor to consider in choosing a gouge is the shape and length of the handle. If you're going to carve for long periods at a time, you will want to choose a gouge that fits comfortably in your hand. "Palm" tools that have a wider base will keep your hand from feeling cramped. The length of the tool shaft is not usually important in carving gourds unless you choose a gourd shape that has a very tight curve, thereby re-

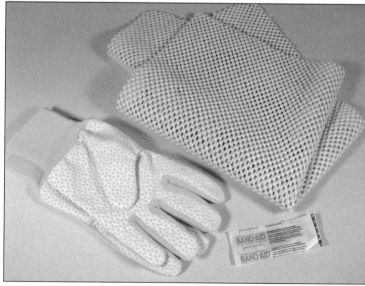

**Supplies used for hand and lap protection
— keep band-aids handy!**

You will need a few marking tools: a pencil, good quality eraser, flexible ruler or measuring tape, and a compass. An ice-pick or awl is useful for making hanging and ventilation holes. Regular garden clippers will work better than household scissors for trimming gourd stems to an artistic length. Last but not least, wood putty or plastic wood can be used to fill in a "chip" that goes too deep and leaves a hole on the gourd, but hopefully you won't need it! Cutting and staining supplies will be introduced in detail in the bowl-making project.

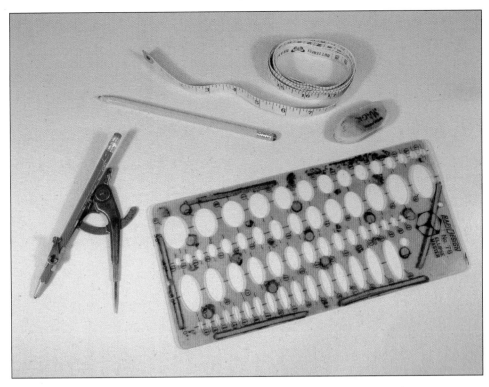

Marking tools: pencil, eraser, measuring tape, compass, plastic flexible ruler for drawing lines on curved gourd.

While these tools are useful, they're not required.

Making Chips!

Choose a sturdy medium-to-large size whole gourd to begin your practice. This may seem like a waste of a gourd, but practicing on a low quality gourd or scrap pieces of gourd will just be frustrating. A thin gourd will break. Exerting pressure on curved scraps of gourd will cause the pieces to break and also puts your holding hand dangerously closer to the gouge. Holding the gourd in your lap will be more comfortable than using a table. Slight pressure from your knees will steady the gourd with control exerted from your non-carving hand.

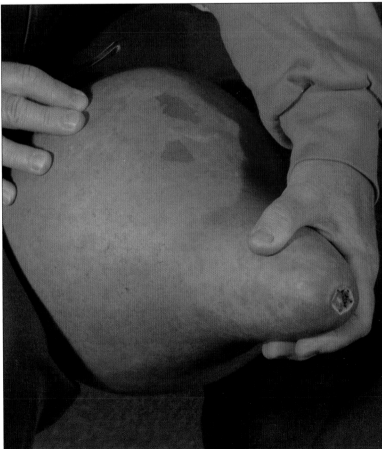

This is a good position for holding a gourd.

The basic chip used in all the patterns in this book consists of two parts — the stop cut and the chip. The "stop cut" is made by placing the gouge at right angles to the gourd surface and pushing it into the surface with a little side-to-side wiggle until a thin "U" is cut in the gourd. Next, begin the "chip" by sliding the gouge back from the stop cut and lowering the angle of the gouge to about 45 degrees; dig lightly into the gourd, and with a side-to-side motion of your wrist, push the gouge forward. It will take some time to find the right depth of carving: too shallow — the gouge will skip off the surface of the gourd, and too deep — the gouge will be very hard to advance and may make a hole all the way through the gourd shell. Continue the chip until it meets the stop cut and the chip will pop out leaving a nice rounded edge. The thin tips of the U will extend beyond the chip.

That's it! Really, that's all there is to making chips with a U-shaped gouge!

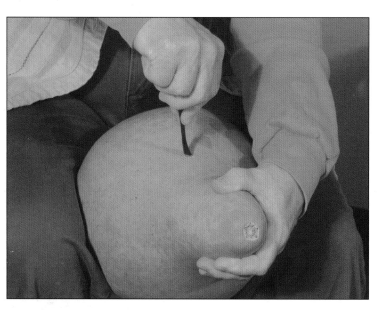

Recommended hand positioning for making a stop cut

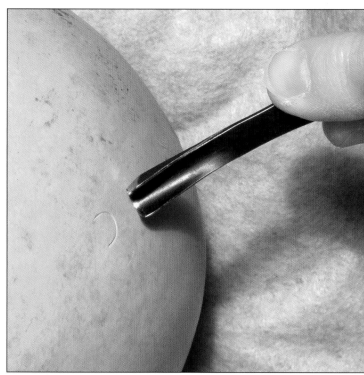

U-shaped stop cut. The gouge is pulled back, ready to start chipping.

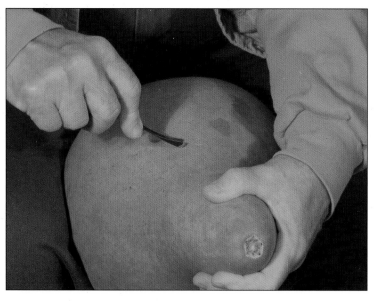

Notice the hand positioning for making chips.

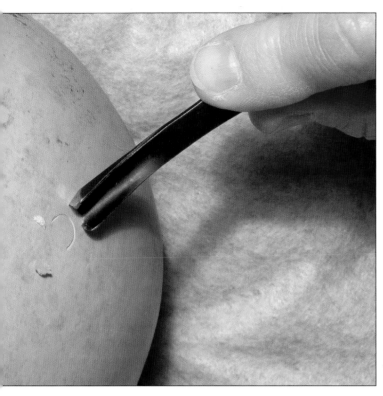

Chipping in progress. Notice how the rocking motion of the gouge makes the shell flake off in small pieces.

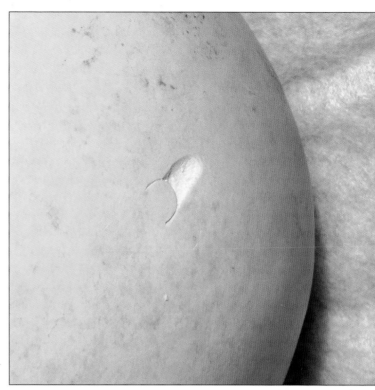

Finished chip

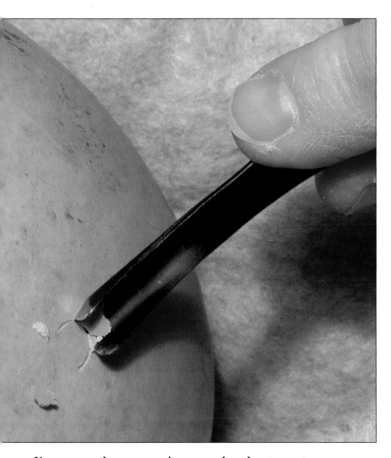

You can see the gouge as it approaches the stop cut.

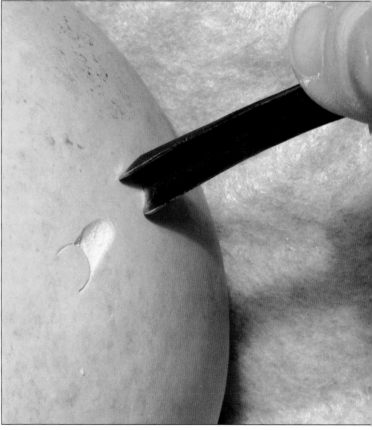

Make the stop cut for a second chip behind the first one.

Although the width of the chip is determined by the gouge, you can vary the length of the chip depending on how far behind the stop cut you begin to carve. Try making some different length chips. Short chips are easy. Longer chips are harder to carve evenly, but practice will help.

Shown are chips of different lengths.

Now try making a row of evenly sized chips. Draw a pencil guideline on the gourd and make a set of pencil spacing marks about an inch apart. Make the stop cuts on the spacing marks, and make each chip about ¾ of an inch long.

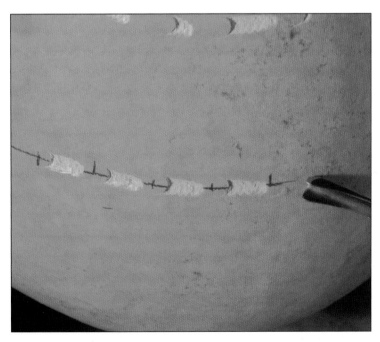

A row of uniform chips spaced 1" apart

Most patterns in the book are based on rows of carving, with each row carved in the opposite direction. To start a second practice row, turn the gourd upside down. Make the top tip of the stop cut in this second row touch the bottom of a stop cut in the first row of carving. Continue making new stop cuts that join with the first row. You will see that the length of the chips – which must fit between the stop cuts — is fixed by the spacing of the first row of carving.

The key to neat carving is this "interlocking" of stop cuts between adjacent rows of carving. Although the stop cuts are barely visible before staining, they will pick up the leather dye and appear as thin graceful lines that separate and define the rows of carving. The stain will also turn the chip dark and that will make the gentle light brown background S-shaped swirls much easier to see.

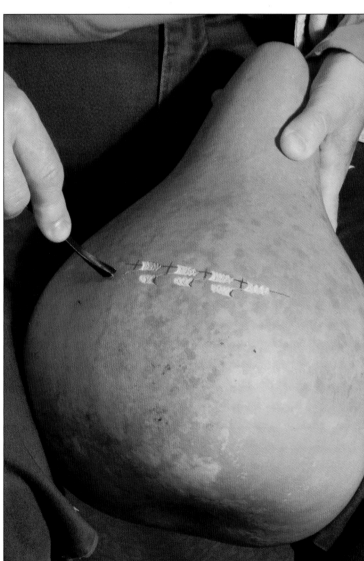

Second row of carving shows interlocked stop cuts.

For a fixed size gouge, there are two variables that will affect the appearance of your carving. The first is how far apart the chips are spaced. In the practice row of carving, you spaced the stop cuts on marks that were 1" apart, which will produce very loose "S" curves. If the marks are ¾" apart, the curves will be tighter. At ½", the curves will be very tight.

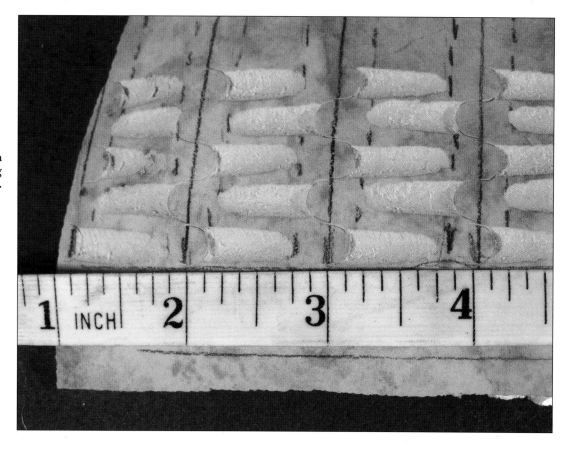

Guidelines show a band of chips being spaced 1" apart.

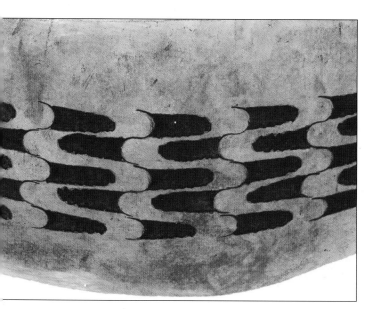

Same 1" spacing, but stained to bring out the "S" curves

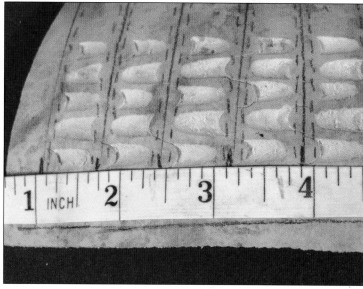

These guidelines show a band of chips spaced ¾" apart.

Same ¾" spacing, but stained to bring out the "S" curves

The second variable is the actual length of the chip between the stop cuts. In the previous examples, each chip filled about ¾ of the space allotted. In the next set of sample carvings, the stop cuts are all spaced ¾" apart, but the chip fills ¾, ½, and ¼ of the space between the stop cuts, giving a very different appearance to the curves when stained.

For most of the projects in this book, the spacing marks for the first row of carving will be about an inch apart and the chips will cover about ¾ of the allotted space between stop cuts. The pattern layout directions will specify where to draw the guideline for the first row of carving — usually on the widest part of the gourd that is to be carved. So it will follow that if the first set of chips carved is the largest, and if each successive row of carving is fixed by the size of chips in that first row, each new row will have shorter chips and will therefore be easier to carve!

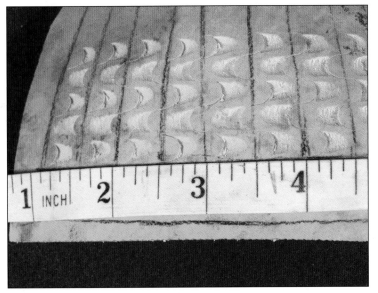

Guidelines that show a band of chips spaced ½" apart

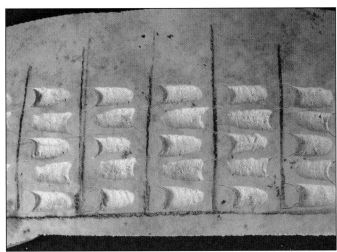

Guidelines spaced ¾" apart with chip taking up ¾ of the space between stop cuts

Same ½" spacing, but stained to bring out the "S" curves

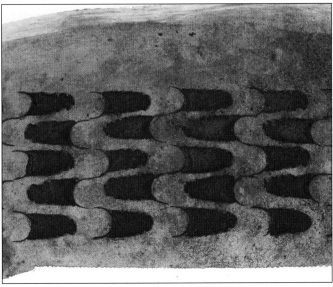

¾ of the space used, but stained to bring out the "S" curves

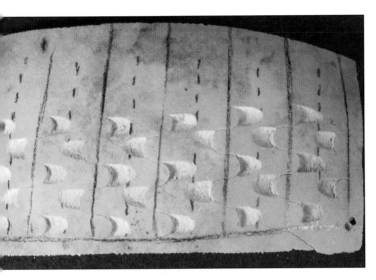

Guidelines spaced ¾" apart with chip taking up ½ of the space between stop cuts

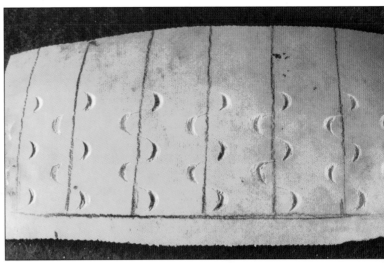

Guidelines spaced ¾" apart with chip taking up ¼ of the space between stop cuts

½ of the space used, but stained to bring out the "S" curves

¼ of the space used, but stained to bring out the "S" curves

Chapter Two:

Carving Band Patterns

A Band Pattern on a Bowl

A good first carving project is a simple band pattern on a bowl. Kettle gourds are readily available and are a natural choice for making bowls because of their flat bottoms and gently sloping sides. The width of kettles varies from about 4 to 12 inches at the widest part. They are sometimes referred to as "birdhouse gourds" since the large kettles are commonly used as purple martin houses. Set the prospective bowl gourd on the floor and look down from the stem end to make sure it's nice and round.

The kettle gourd shown in this project is about 6 ½ inches across at the widest part. Mark a pencil line on the gourd where the top will be cut off to make the bowl. A simple marking device can be made with a dowel rod inserted into a flat wooden base; a small pencil holder slides up and down the dowel and is held in position with a setscrew. (A stack of books and magazines with a pencil laid flat on top will also work.) Hold the pencil against the gourd and turn the gourd slowly to make the cutting guideline. Now draw a second line at the widest part of the gourd where you want to begin the bottom row of carving. In general it's easier to carve the gourd bowl before actually cutting the top off since there is more room to hold the gourd securely by gripping the stem end, allowing you to keep your hand safely away from the carving gouge.

Bowl project

Shown are an assortment of kettle gourds.

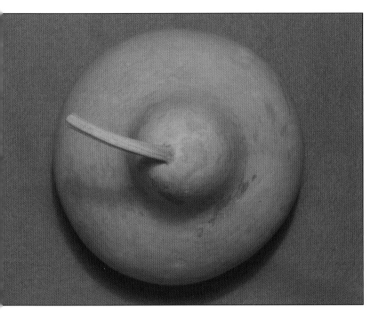

Check for "roundness."

Cut line for bowl.

Marking tool, and now ...

Shown is the guideline for bottom row of carving.

Mark the spacing for the bottom row of chips on the lower line, fudging the last few marks if necessary to avoid having the last chip in the row be too obviously small. You can use a ruler or tape measure held against the gourd, but setting compass points the desired distance apart is a quick way to make equal spacing marks. Remember that this will be the row that has the longest chips, so choose a comfortable size space (about ¾" in this example), knowing that the chips will get slightly smaller with each upward row, depending on how gently the sides of the gourd slopes.

You will find your own best way to hold the gourd while carving, but bracing the gourd between your knees in your lap and holding the top of the gourd with your non-carving hand works well for the first row. When carving the second row, the gourd must be turned upside down and can be wedged in your lap with the biggest part between your knees. Your non-carving hand is more likely to be closer to the gouge when you carve in this direction, so carve the second row cautiously. Be sure to interlock the stop cuts of the second row with those of the first. If you make two stop cuts next to each other in the row, it will be clear to you where the chip must go. Continue to carve the second row of chips around the gourd.

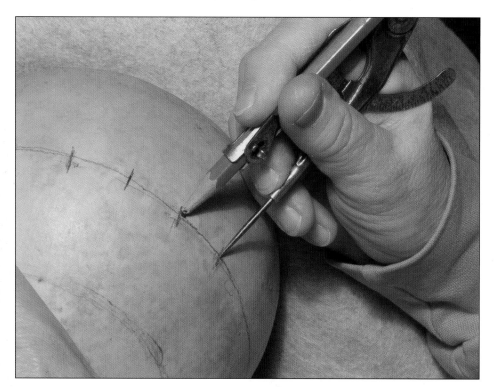

Mark the chip length.

Bottom row of carving

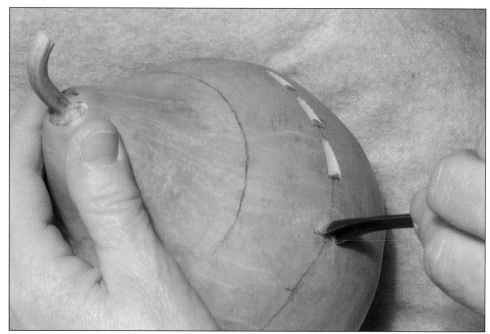

For each new row, reverse the direction of the carving by flipping the gourd top to bottom, and continue until you have carved a band of a pleasing width for the size gourd you have chosen. Leaving an inch or so of space between the top row of carving and the cut line will give the edge of the bowl a nice look and will allow room to add a rim treatment later if desired.

Whether you stain before or after cutting the top off will affect the color of the cut rim. For the most options, stain the gourd first before cutting, allowing the cut edge of the rim to retain its natural color of the interior of the gourd. Any kind of stain and finish recommended for wood can be used on gourds. A combination of leather dye and brown paste shoe polish works well to heighten the contrast between the chips and the gourd shell. The alcohol-based leather dye dries quickly, and the shoe polish selectively darkens the carving and gives a warm glow when buffed.

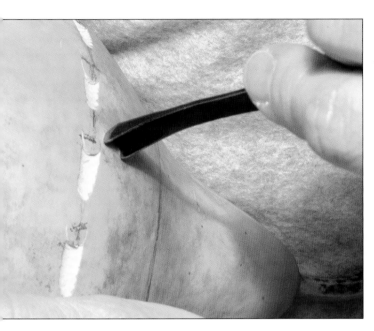

Interlock stop cut for second row, and ...

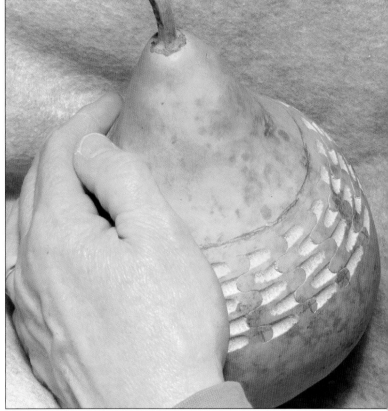

Fifth row finished

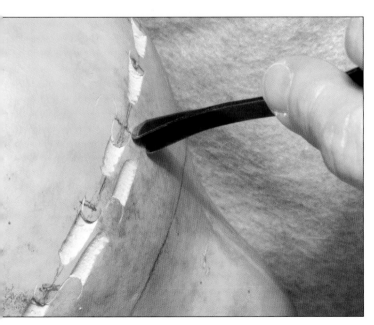

Carve second row.

Leather dye comes in dozens of colors, but all the projects in this book are stained with a medium brown to give a natural color. Each gourd takes the stain a little differently, and the longer the stain is left on, the darker the color. You can add a little rubbing alcohol to the full strength dye before application to lighten the color. Trying to lighten the color of a stained gourd does not usually give satisfactory results.

Cover the work area with newspapers, and use rubber gloves to keep the stain off your hands. Either put the small bottle of dye in a cup to prevent upsetting the bottle, or pour a small amount of dye in a cup, so you don't spill the dye while using the applicator that comes with the dye. Remove the pencil guidelines before staining since they will sometimes show through. A regular pencil eraser is fine for small areas, but a damp rag will be quicker to use on large areas. You can leave the pencil cut line, since it will disappear when you make the cut.

Shown is a small section of carving after being stained.

Staining supplies: brown leather dye, alcohol thinner, daubers for applying dye, soft cloths for blending dye, latex gloves.

Shake or stir the dye before using and occasionally during the staining of a large gourd. Apply the dye quickly to about ¼ of the carved area, and wipe it off immediately with an old rag, blending with a circular motion so that no dark edges are left. Since the dye is very liquid, check for drips under the gourd, and use that excess dye to color the non-carved portions of the gourd. As you work around the gourd, staining in small sections, blend from one section to the next, and be sure the color extends evenly over the rough area at the bottom of the gourd (where the blossom was attached) and also well above the cut line. If you see you have missed a spot, dab on a bit more dye and blend.

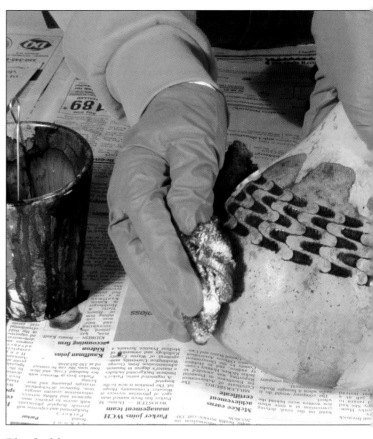

Blend with rag.

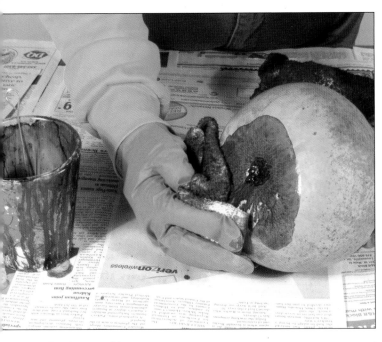

Stain the gourd bottom.

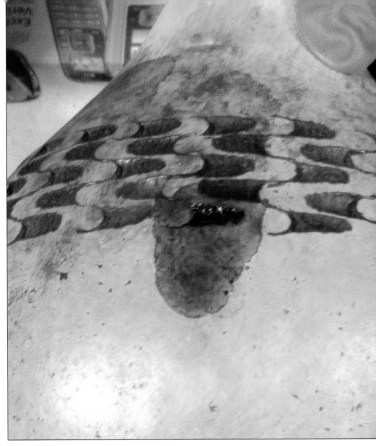

Reapply stain.

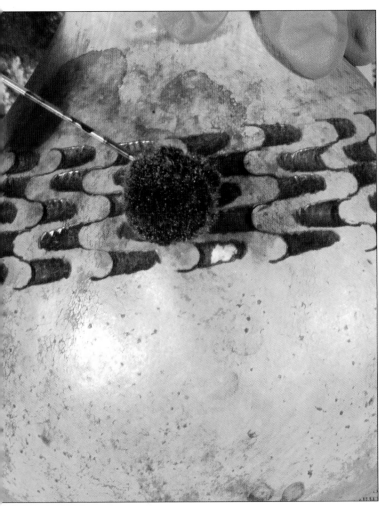

Stain missed this chip, so ...

Missed spot retouched

Since the leather dye dries so quickly, a paste shoe polish finish can be applied immediately. Use an old toothbrush to work brown paste shoe polish into the chips to darken the carving; the un-carved areas will be darkened to a much lesser degree. A regular shoe polish brush can be used to apply the polish quickly to large un-carved portions of the gourd. You can apply polish to the whole gourd at one time. A soft terry cloth rag is good for removing extra polish and buffing the gourd to a natural shine.

Polishing supplies: brown paste shoe polish, toothbrush to work polish into carving, regular shoe polish brush for uncarved areas, soft cloths or shoe brush for buffing.

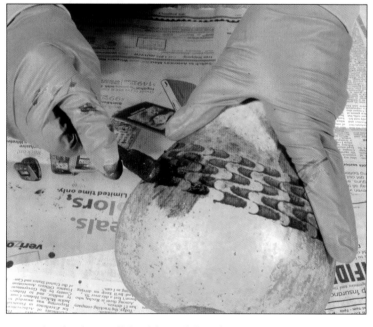

Apply shoe polish with tooth brush.

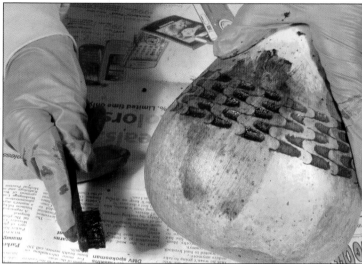

Brown polish darkens the carving.

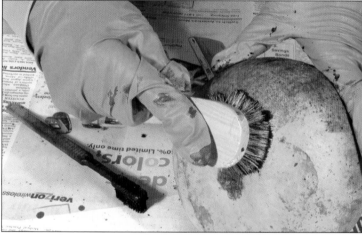

Use a larger brush for applying polish to uncarved areas, and ...

Buff to a nice shine.

For the actual cutting of the bowl, use a tool that you are comfortable with, remembering that you should be able to hold the gourd securely with one hand while using the saw with the other hand since gourds are hard to put in any kind of vice. If you have never cut gourds before, an inexpensive and easy-to-use tool is a "keyhole" saw. While cutting and sanding the bowl, use a dust mask if you have dust allergies.

(Using a keyhole saw does not make as much fine dust as power tools, but better not take chances if you are working indoors. Also some gourds are "dustier" than others when you clean out the insides.) Insert the pointed tip of the keyhole saw on the pencil cut line and begin sawing, slowing down when you almost complete the cut around the gourd so that it does not crack.

Keyhole saw and dust mask

Insert the saw on the cut line.

Finish the cut, and ...

Scoop out the seeds and any papery material, and sand the cut area of the rim with fine sandpaper — be careful not to mar the finish by letting the sandpaper touch the outside of the bowl. Sand the inside of the bowl. Any dust that settles in the carving can be removed with a soft shoe brush. Use a tack cloth to remove dust from the inside of the bowl. The inside may be left natural if the bowl is just used for display. Use a food-safe product like woodworkers use on salad bowls if you want to serve something like chips from your bowl. Gourd bowls do not make satisfactory containers for moist contents.

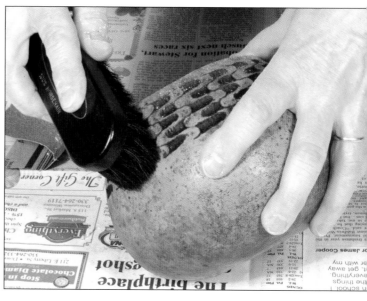

Remove dust from chips.

Clean out the seeds.

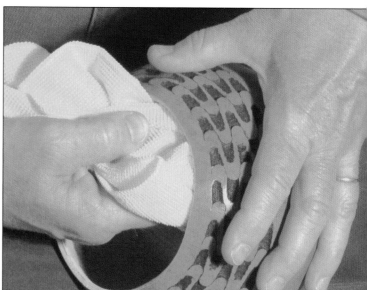

A tack cloth is used to remove dust from inside the bowl.

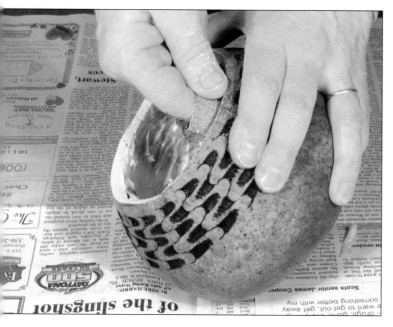

Sand the rim and the inside.

The bowl is finished.

28

Varying the "Band" on Bowls

There are many variations on this basic band pattern. Just increasing or decreasing the number of rows of carving will give a heavier or more delicate look to the bowl. Or you can make two or more rows of carving by skipping several rows of chips. You can separate the band into blocks to give a pleasing effect by skipping a number of chips in each row. You may have to do some estimation of "chips per inch" to make sure the blocks come out even; however do not stress over this since you will not see all sides of the bowls at one time, and a small variation will not be noticeable.

Triangle Pattern

A triangle pattern is carved from the top down, with a guideline for the first row of carving placed close to the cut line. Although the chips will get larger in the second row, you are going to be tapering the chips to fit your triangles, so this is not going to be a problem. As you carve rows downward, leave out a few chips in each new row, and decrease the size of the chips. If you're feeling adventuresome, this pattern can also be reversed to add triangles of carving going upward.

A double row of carving

Triangle pattern

Rows of carving split vertically

Shown are triangles carved down from top and then up from bottom.

29

Curves

As a last example in carving bands on bowls, here is a bowl with a curved rim and a curved band. This bowl will be cut before carving so that the carving will be spaced evenly from the rim. Be sure to choose a very sturdy gourd for this project, since the pressure of the carving will be concentrated on the cut rim area instead of being spread out over the whole gourd. The rim of this bowl will be cut into a gentle curve along a freehand pencil line. After cutting and cleaning the inside of the bowl, a compass can be ran along the edge to make the guideline for the top row of carving. In the same manner, make a guideline for the bottom row of carving, and add a line half way between these two lines. Begin carving on this centerline and carve rows both upward and downward until you reach the lines that mark the boundaries of the area set for the band of carving. Mark your spacing lines slightly smaller than usual because the width of chips in rows both up and down will be influenced by the curvature of the line. As you carve this bowl you may notice that your non-carving hand feels very cramped while gripping the gourd by the bowl's edge — this is another reason for opting to carve a bowl before cutting when possible.

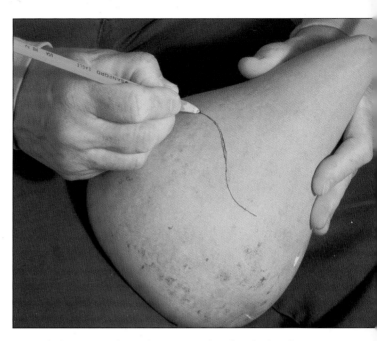

Doing it freehand, draw a curve cut line for the bowl.

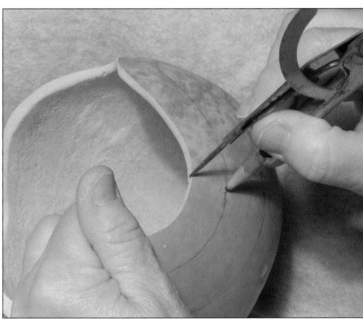

Use a compass to copy curve.

Curved edge bowl project

Staining

As you stain this bowl that was cut first, it will be difficult to get the color to spread evenly to the top of the bowl without getting the dye on the rim. Solutions to this problem are to re-sand the rim where needed, or to stain the entire rim and interior dark. If preferred, the rim and interior can be painted with acrylic paint. Staining and/or painting the inside of the bowl can also solve the problem of any "bleed-through" of the leather dye from deep carving, and it can cover up a blotchy interior due to mold mosaic.

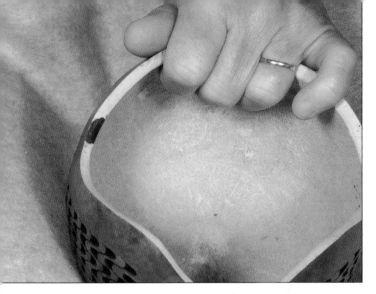

Shown is a stain that accidentally seeped onto rim.

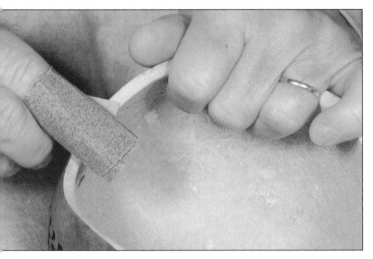

Sanding out the stain and voila ...

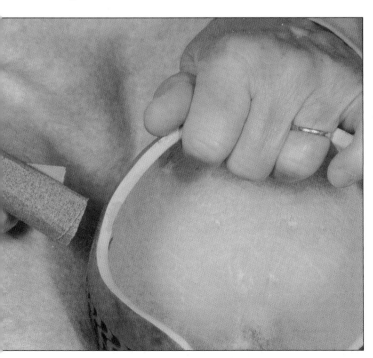

Stain gone!

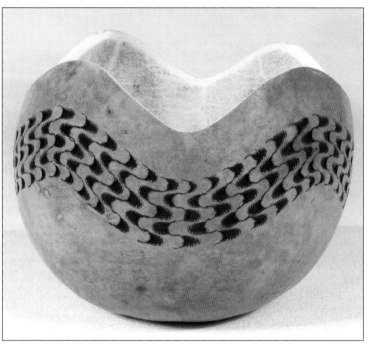

The curved edge bowl is finished.

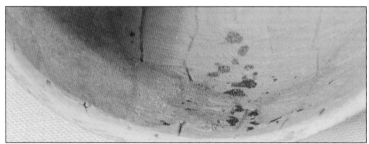

An example of a stain that seeped through when the outside surface was stained

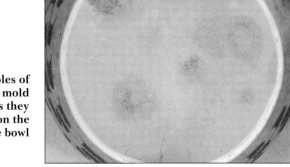

Examples of natural mold patterns as they appear on the inside of the bowl

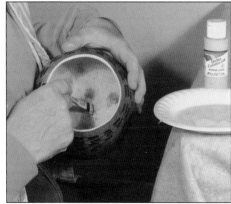

Paint the inside with light color acrylic paint.

A Band Pattern on a Birdhouse

Carving a band pattern on a kettle gourd to make a usable wren house will be very similar to the bowl project, but will involve more rows of carving as well as some new construction details related to the birdhouse. Position the entrance hole in the center of the body of the gourd by drawing around a quarter. When cut on the pencil line, the hole will be 1" in diameter, which is large enough for a house wren, but small enough to exclude pest birds like English sparrows. The kettle gourd for this project is about 6 inches across at the widest point — any cavity 4 inches or larger is enough space for a house wren. Using a compass with its point in the center of the entrance hole, draw a circle about ¼" larger to remind you not to carve too close to the entrance hole.

Wren house project

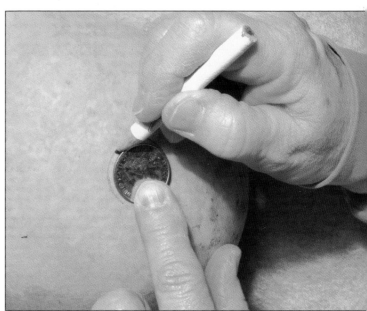

Mark the entrance hole, and ...

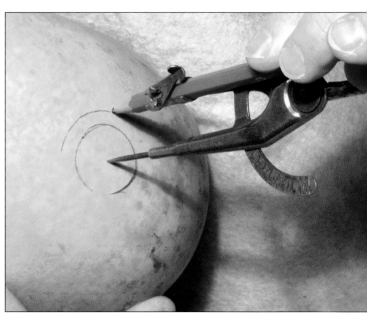

Draw a guideline around entrance hole.

Using the level line-drawing tool, draw a pencil guideline around the widest part of the gourd. Start your spacing marks on this guideline even with the center of the entrance hole and continue the marks about half way around in either direction so that any "fudging" of chip lengths will be in the back of the birdhouse. Begin carving rows upward from the guideline, and when you reach the level of the hole, carve only full-length chips that are outside the pencil circle around the entrance hole. Then go back and carve shorter chips around the hole using the outer pencil circle as a guide until you have a pleasing border around the hole.

Continue carving upward rows until you have most of the gourd covered. Then go back to the first row of carving and carve some more rows below your first row of chips. You will see that these chips will also be shorter and easier to carve than the first row, illustrating why it is important to put the pencil guideline for the first row of carving on the widest part of the gourd.

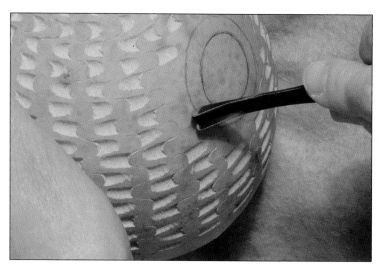

Fill in "short" chips up to the guideline around the entrance.

To cut the entrance hole, insert the blade of the keyhole saw into the center of the hole and rough out the hole to within about 1/16" of the pencil line. Roll a tube of medium grit sandpaper around your finger and smooth the edge of the hole up to the pencil line by turning your finger and/or the gourd. If a quarter just hangs up in the hole, it is the proper size. Shake out the seeds and papery material through the entrance hole. If the seeds seem to be stuck, loosen them up by poking the insides with a screwdriver or table knife, but be careful not to exert pressure on the sides of the hole to avoid nicks or breaks at the hole edge. It isn't necessary to remove every bit of inside material since the birds will remodel to suit, however any loose material will shake out as you stain and make a mess.

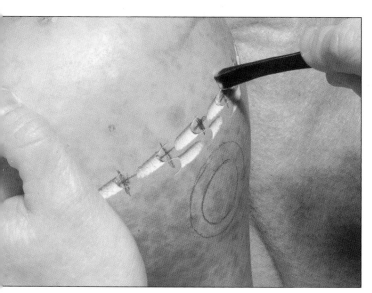

Interlock the stop cuts for third row of carving, and ...

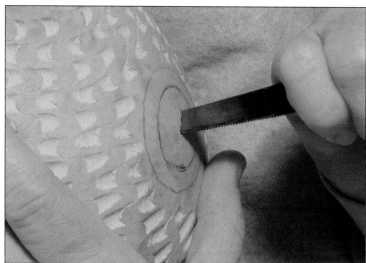

Insert the keyhole saw to cut the entrance hole.

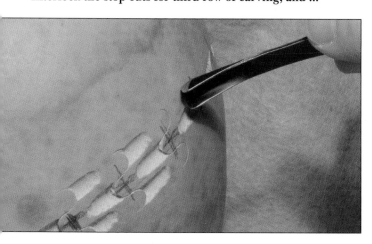

Carve a row below the first row.

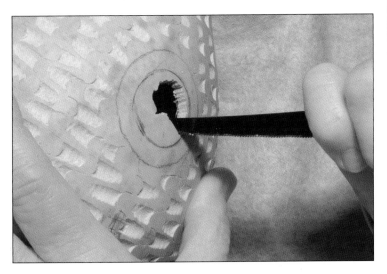

Rough out the entrance hole.

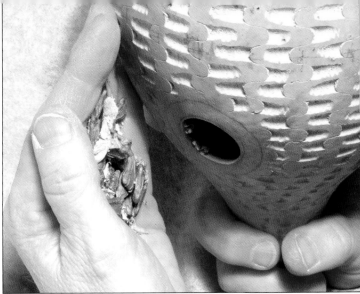

Shake out the seeds ...

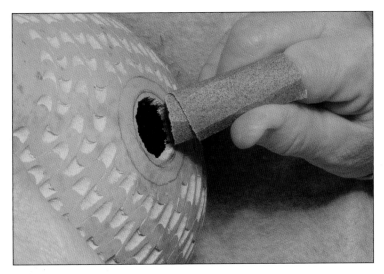

Ready to sand

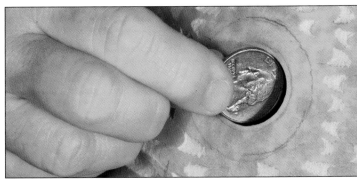

Check for proper size with a quarter ...

Use an awl or ice-pick to make two holes near the top of the gourd for hanging, and several holes in the bottom of the gourd to promote good air circulation. These bottom holes are usually referred to as "drain holes," but unless there is a sideways rain, water in a gourd birdhouse is not often a problem. (And to ease your mind even more, the cup of fine grasses that holds the wren eggs and young nestlings is built on top of a substantial platform of cross-laid twigs.)

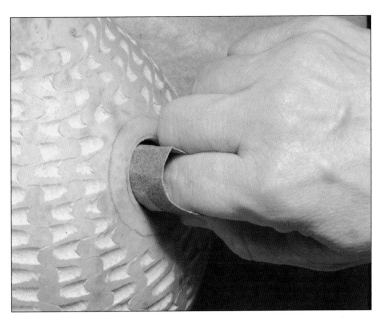

Twisting the sandpaper around your finger can help make the edge smooth.

Insert the ice-pick at right angles to the gourd ...

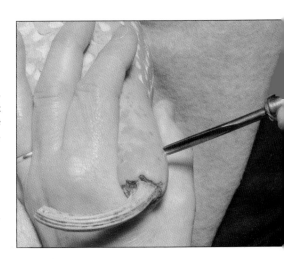

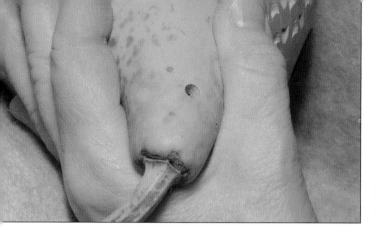

Make a hanging hole on each side ...

Stain the front section including the rim of the hole.

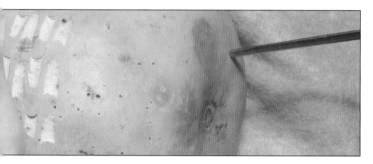

Insert the ice-pick to make holes in the bottom of the birdhouse ...

Four to five holes should provide good ventilation.

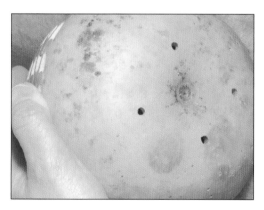

Stain the top section including the stem.

Remove all pencil lines and stain in small sections for uniform coverage, paying particular attention to the inside rim of the hole and the stem. Apply the wax shoe polish finish, which will also help to waterproof an outdoor birdhouse.

Erase pencil lines, and ...

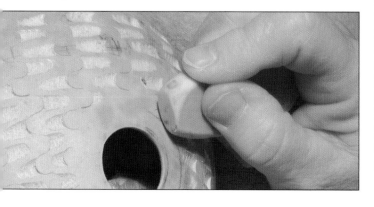

A simple leather lace makes an excellent hanging loop. A single package of leather laces for work boots should make loops for 6-8 birdhouses. Many craft stores sell spools of leather lace material if you plan to make lots of birdhouses. Use cellophane tape to attach the lace to one end of a straightened-out jumbo paper clip. Then use the flexible paper clip like a needle to guide the lace through the hanging holes. Hang your wren house where you can easily see it to enjoy the nesting birds and to hear the wren's cheerful bubbly song. House wrens are birds that do not mind being close to people, so a porch overhang is a good location. Chickadees may make an early first nesting in a wren house, but when the migratory wrens arrive, the wrens will usually take over the birdhouse once the chickadees have fledged.

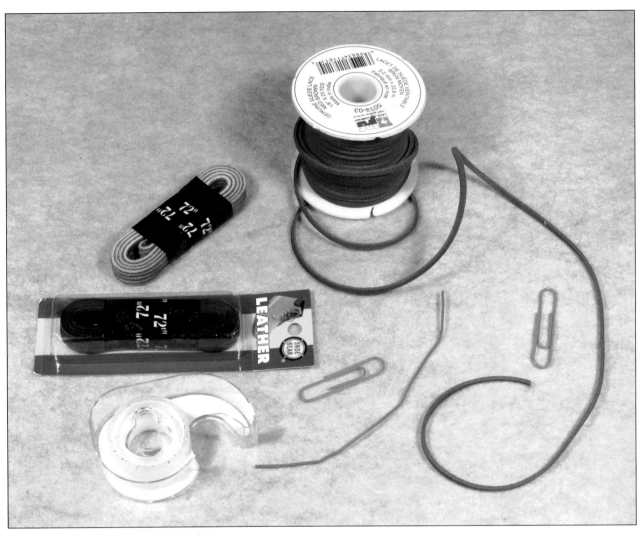

Shown are the supplies for making a hanging loop.

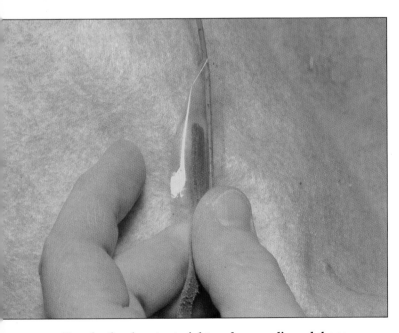

Tape leather lace to straightened paper clip and then ...

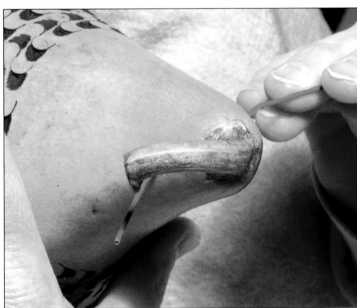

Thread the lace through the hanging holes and ...

Finished!

Smaller band of carving

Lower band of carving

An even lower band of carving

Variations of the Band Pattern

The width of the carved bands as well as the placement of rows of carving can add variety to this simple band pattern. Birds are attracted to birdhouses based on the size of the entrance hole, according to the theory that a bird selects an entrance hole just slightly larger than its body size; this instinct prevents a larger predatory bird from following it into the house. For bluebirds, the entrance hole should be about 1½" in diameter in a gourd with 6-8 inches of cavity space. Tree swallows and tufted titmice will also use this size birdhouse entrance. However since English sparrows will fit in this size hole, it may be necessary to clean out their nests repeatedly to discourage them.

A carved birdhouse will weather well outdoors, especially if taken in during the winter season when no birds are nesting. Although the finish may dull, another coat of shoe polish can easily be applied to bring back the shine when the birdhouse is taken down in the fall. And, yes, the old nest can be shaken out through the entrance hole. If necessary, poke the nest contents with a screwdriver or table knife to loosen the twigs and grass. This is a very dusty job best done outdoors with the wind at your back!

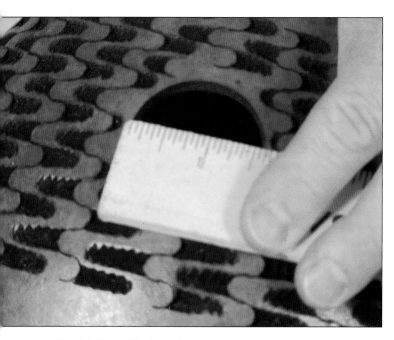

Bluebird sized hole

Size comparison of bluebird house (left) and wren house

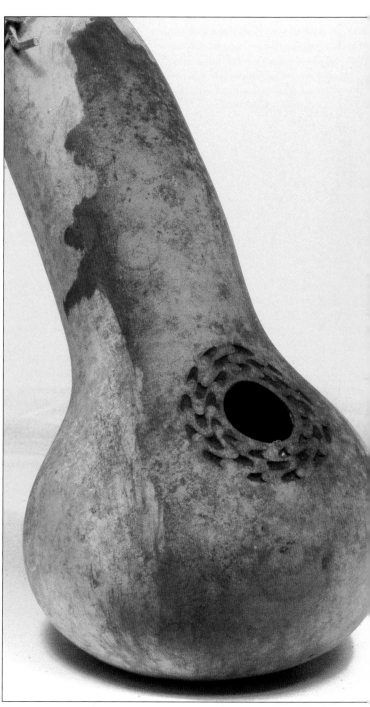

See how a new coat of shoe polish can restore the shine to a weathered wren house!

Chapter Three:

Carving Top to Bottom

Carving a Canteen Gourd

Once you have carved a band around a gourd, there is nothing to stop you from carving a whole gourd top to bottom. The interesting shape of the canteen gourd is perfect for this project. These flat gourds range in size from about 6 to over 12 inches across, with the gourd selected for this project measuring about 7 inches.

Canteen project

Canteen gourds

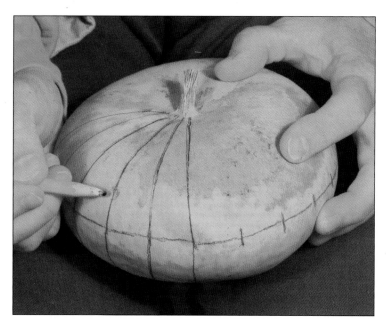

Carving guidelines

Start with a level guide line around the widest part of the gourd, and add 1-inch marks for spacing the first row of chips. Think about what will happen as you carve rows of chips toward the stem end of the gourd. The chips will get shorter and shorter until eventually there will not be enough room for all the chips. The pencil lines drawn to the top of the gourd are not really necessary since the size of the chips in each subsequent row is fixed after the first row is carved, but the lines clearly show what will happen to the chip size as the carving reaches the stem. When the spacing gets tight, just omit a chip at random. Don't stress over which chip to leave out — there are no rules, just do it! (Well maybe one suggestion would be not to skip more than 2-3 chips in any one row.) When the gourd is stained, the missing chips will not be noticeable to the casual viewer who sees the overall pattern, and in fact they are hard to find even when you look for them.

Pencil arrows show omitted chips.

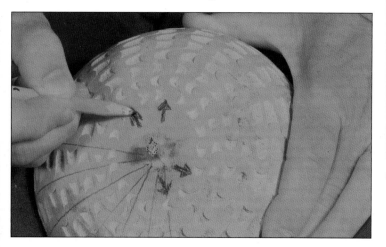

Continue carving right up to the stem, skipping one or two chips as needed per row. In the row of chips next to the stem, there are only 10 chips, down from 19 chips in the first row of carving. Now carve rows toward the blossom end of the gourd. Because of the large blossom scar, not as many chips need to be skipped on the bottom, resulting in 16 chips in the innermost row. The blossom scar cannot be carved – it is a very crumbly texture.

Carving complete to stem

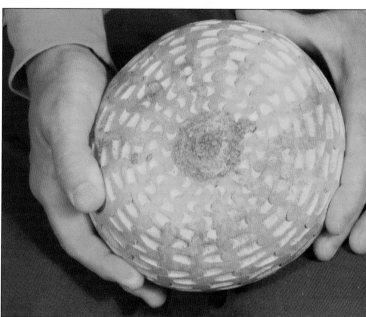

Carving complete to blossom scar

Finished canteen

Narrow Un-carved Band

A simple variation on this pattern leaves a narrow band of un-carved gourd. And if desired, cutting through the middle of the band produces a bowl with a lid. The keyhole saw used to cut the previous bowl will remove shell equal to the width of the blade. For a lidded bowl, it's best to switch to a cutting tool with a thinner blade so long as the blade is sturdy enough to cut the gourd. The blade should be inserted at a slant, making a lip for the lid to set on. If the saw is inserted straight up and down, the lid will fall into the bowl.

In general a lidded bowl will look best if the lip of the bowl is stained dark, making the cut line appear narrower and more uniform than a natural light color cut line. The interior of the bowl is also stained dark so that it does not draw the eye away from the carving on the outside.

Thin craft knife for cutting lidded bowl

Cutting the lid

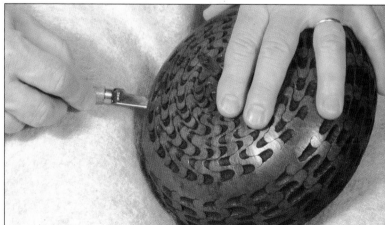

Finished canteen, bottom view

Carving Bottle Gourds

Most types of gourds can be carved in this fashion from top to bottom, but a few variations of bottle gourds present some extra challenges. Bottle gourds with short necks are usually called Chinese or Mexican bottles. They range in size from giant bottles up to 24 inches tall to mini bottles (often called "jewelry gourds" because of their use as earrings and necklaces) about 1" tall. The bottle shown in this project is 10 inches tall and shows an interesting gradation of chip size: largest around the middle of the lower lobe, decreasing dramatically at the neck, and then flaring out a little before coiling tightly around the stem.

As is standard for a band pattern, draw the guideline for the beginning row of carving around the widest part of the bottom lobe. Mark your comfortable chip spacing size of about 1". The vertical pencil lines are not necessary, but will give you an idea of how the chips will look as the gourd slopes in and out. A problem will occur if the lobes are placed so close together that there is not enough room for your hand if you are using a palm gouge. Choose your gourd carefully so that the palm gouge may be used freely — or switch to a longer handled gouge.

Finished, lid in place

The inside is stained dark.

Chinese bottle shapes

Pencil guidelines show where the chips are placed.

Bottles with long necks

Finished Chinese bottle

Long necked bottles may reach a size of 30 inches and are "goose" gourds if the top lobe is elongated and "Indonesian bottles" if the top lobe is nicely rounded. A flat, narrow necked bottle is a "siphon." When bottles are grown hanging on a trellis, the weight of the bottom lobe makes the neck straight; bottles grown on the ground have graceful curves.

If a long-necked bottle gourd has much of a curve to the neck, then there will need to be more rows of carving on the outer side on the neck than on the inside. One way to approach this problem is to start with guidelines around the middle of the bottom lobe, around the middle of the top lobe, and about half way up the neck. Make 1-inch space lines on the line on the bottom lobe and carve the bottom lobe down to the blossom scar, and then carve up to where the neck begins to curve, omitting chips until you have a chip length of about ¾" at the neck. Next, mark 1-inch space lines on the guideline on the top lobe and carve the top lobe up to the stem and then down to where the neck begins to curve. Mark ¾" spacer lines on this line around the neck, and carve rows on either side of this pencil line until the rows of carving meet on the under side of the neck.

Then fill in rows on the outer part of the neck, continuing each row until it blends into the existing rows. Interlock stop cuts when possible. It may look awkward while you are carving, but when stained, the extra chips will not be noticeable in the overall pattern.

Three pencil guidelines

Finished Indonesian bottle

**Finished
Indonesian bottle,
second view**

Chapter Four:

Carving in Circles

Carving a Circular
Pattern on a Birdhouse

Round gourds inspire circular designs! Naming round gourds is fun — there are bushels, basketballs, and cannonballs down to ornament size and even smaller. To begin this wren house on a gourd about 9 inches across, center the quarter to draw the entrance hole. Use a compass to draw a guideline for the outside of the carved circle. Make the circle as big as possible, but leave just enough space to have an un-carved border that looks symmetrical around the carving.

**An example of a circular
pattern on a birdhouse**

Here's an assortment of round hardshell gourds.

With a flexible ruler or a piece of light weight cardboard, draw horizontal and vertical lines through the center of the circle to the outer circle guideline. Then continue dividing these pie-shape wedges in half until you've reached a comfortable length of chip to carve on the outer circle guideline. With the stop cut on the spacer line, carve the chip hugging the circular line, and continue on around the circle for the first "row." These chips will have a very slight curve that will make the whole carved area look truly round. The second row will be carved on the inside of the first row by reversing the direction of carving and interlocking the stop cuts.

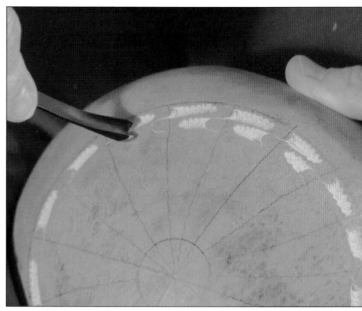

Second row of carving

Continue with each row toward the center of the gourd until there is a pleasing un-carved border to the entrance hole. If your carving is not exactly even (which can sometime result from irregularities in the curvature of the gourd), you can erase the entrance hole line and move it a little to re-center it before cutting.

Finish construction and stain the birdhouse as before. When inserting the hanging loop through two holes made near the top of this relatively flat gourd, you will see the advantage of having the flexible paper clip to thread the lace through the holes.

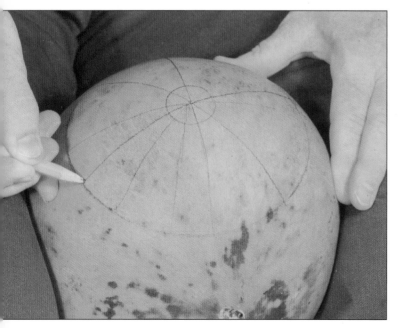

Draw the circular guideline.

Draw the wedges for spacing chips.

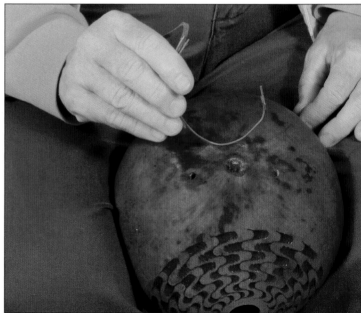

Bent paper clip for threading a hanging loop

Finished birdhouse!

Making Flat Ornaments

With a birdhouse, you do not have to worry about carving into the exact center of the circle, but for other projects, you may want carving that extends all the way to the center. Making a set of ornaments is good practice — and can be done with a gourd that has a soft spot or some other flaw that makes it unsuitable for a whole-gourd project. To avoid a pinhole where the compass point is inserted, use a jar ring, teacup, or small bowl to draw the cutting circle. The open top of these objects allows it to fit tightly on the gourd for easy drawing of the guideline.

Flat ornaments with circular carving

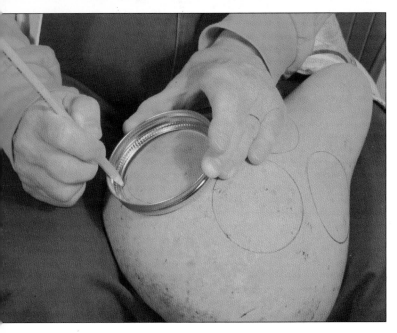

Easy way to draw circles

Divide each space in half.

Divide into thirds, and ...

Because of the curvature of the gourd and small size of these circles, it will be easier (and safer) to carve the ornament before cutting it out. Draw the pie-shaped wedges as before. If dividing the remaining space in half each time results in a chip that is either too small or too large for the outer row of carving, then go back to the circle divided into fourths and divide each fourth into thirds. This does not have to be exact!

Begin carving about ¼" inside the cut line. Depending on the size of your circle, you may have to omit some chips as you have done before from rows near the center. You can experiment and decide how you want the center to look. Maybe not having the center chips meet exactly is OK? If you see you're carving is a little too wide on one side and too narrow on the other, leave a bit of extra space instead of having the stop cuts meet exactly until you are even again. If you look closely at the four finished ornaments in the example, you will see some minor "flaws" — places where the stop cuts don't touch exactly or where the centers are, well, off center, but the overall appearance is fine.

If you really want the center to come out even, you can learn to estimate pretty closely how from the edge to start caving by making a test carving with the desired size circle and gouge. Or if you want it to come out right the first time, you can make a series of stop cuts along one of the straight lines, but this is pretty tedious. (Remember carving is supposed to be fun and relaxing!) After staining the ornaments, glue a loop of ribbon on the backside.

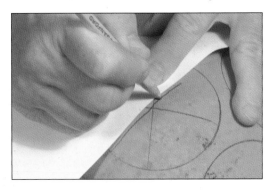

Divide the circles for carving.

Assure a "centered" center.

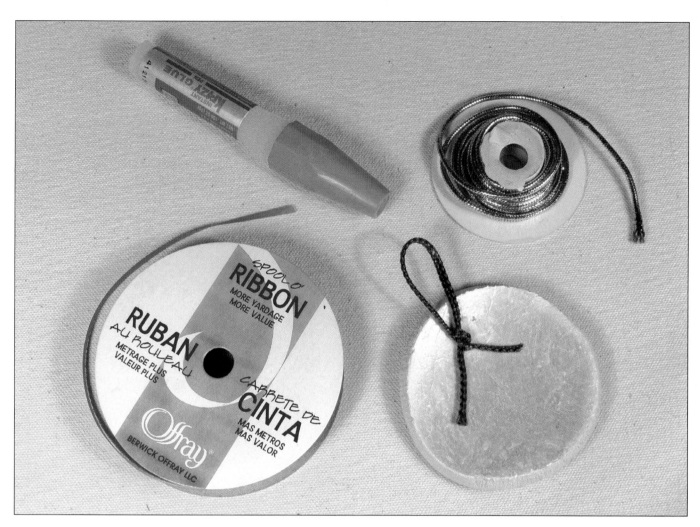

Add the hanging loop.

Carving "Inside Out"

Another option is to begin carving circles from the center outward. Start by making five stop cuts that touch in the center of the circle, and then add a little chip to the backside of each. Then carve outward, interlocking the stop cuts of each row. It takes a little practice to add chips evenly when carving outward. Also it may be harder to get a round appearance on the outermost row of carving unless you carve very evenly.

Carve five stops cuts to make the circle center.

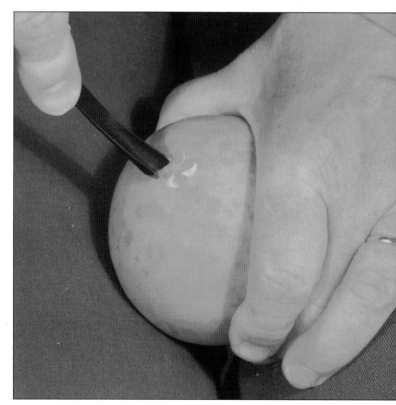

Carving from the inside out will look best on a three-dimensional ornament with 2 or 3 circles of carving. Adding some extra chips to fill in blank spaces between the circles will make the outer row of each circle less well-defined, so it won't matter if the finished circle doesn't come out exactly round. To hang the ornament if a stem is present, just tie a looped ribbon around it; if the stem is missing, pull a thin cord through a pretty bead and super-glue the bead to the spot where the stem was broken off.

Beads and cord for hanging loops.

Finished 3-D ornaments

Chapter Five:

Carving Odd Shapes

Carving a Heart

A combination of band carving and circle carving can be used to carve a heart pattern. This example is made from a kettle gourd about 10 inches across. Use a stencil to make a guideline for the outside edge of the heart. Make chip spacer marks about 1-inch apart on this line. Center a quarter for the wren house entrance hole, and draw a guideline about ¼" to the outside with chip space marks rather close together since you will be carving outward and making longer chips with each row.

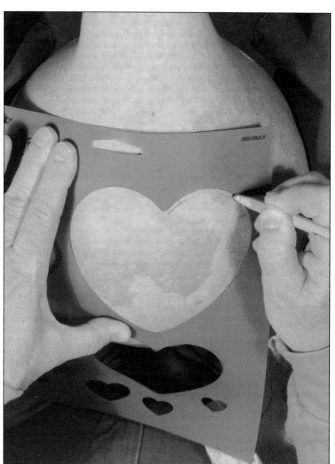

Use a heart stencil, and ...

Heart birdhouse project

Carve a few rows around the outside, and then carve a few rows around the inside. (Just this much carving makes an interesting pattern when stained.) Keep alternating carving rows from the outside and the inside until you end up where the rows touch. There will be three blank spaces leftover; two near the top of the heart and one near the point. Fill in these spaces with random chips, interlocking stop cuts when possible. Finish the birdhouse construction and stain.

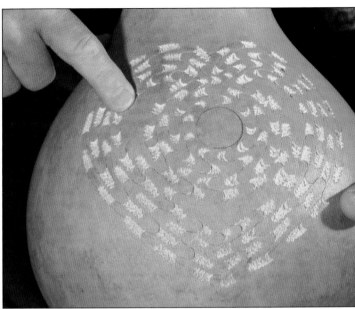

Shown are leftover spaces.

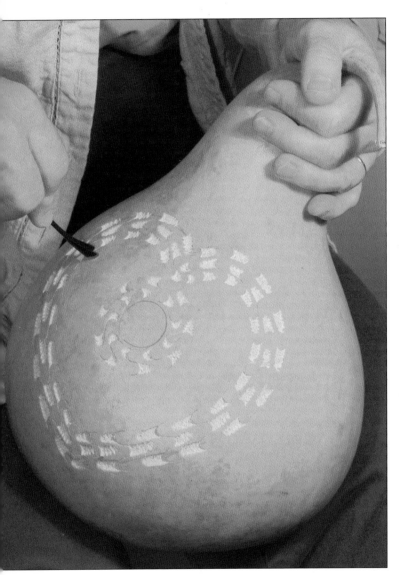

Carve the first few rows.

Finished birdhouse

Carving Apple Gourds

Apple gourds are a relatively new cultivar of gourds and they are tremendously popular. Apple gourds range in size from about 4-8 inches across; they tend to have broader shoulders than bottoms, and may be slightly flattened on the sides. The patterns discussed so far do not look right on apple gourds because the symmetry of the circle or heart seems to emphasize the irregularity of the gourd shape. Even a band of carving does not seem to fit, appearing top heavy since the chips are larger at the top of the band. Apple gourds need a lighter amount of carving that can add decoration without interfering with the overall apple shape. To achieve a lacy outline effect, the stop cuts for this project will be positioned differently.

Apple gourds

Star pattern on birdhouse.

Stencil on the star, and ...

The apple gourd for this project is about 6 inches across. Use a star stencil to make a guideline. To begin this new pattern, make a stop cut with both tips of the cut on the outside of the line. Make a small chip behind the stop cut. Continue on around the star, leaving a small space beside each stop cut. On the inside of the line make a stop cut with both tips touching both tips of the stop cut across the line, enclosing a small oval space between the two stop cuts. Make a short chip behind this stop cut, and then continue on around the star. Erase the pencil line before staining. Since apple gourds are very sturdy, they make excellent birdhouses — a good choice since many apple gourds do not sit evenly on a flat surface.

Position the gouge for carving an inside row.

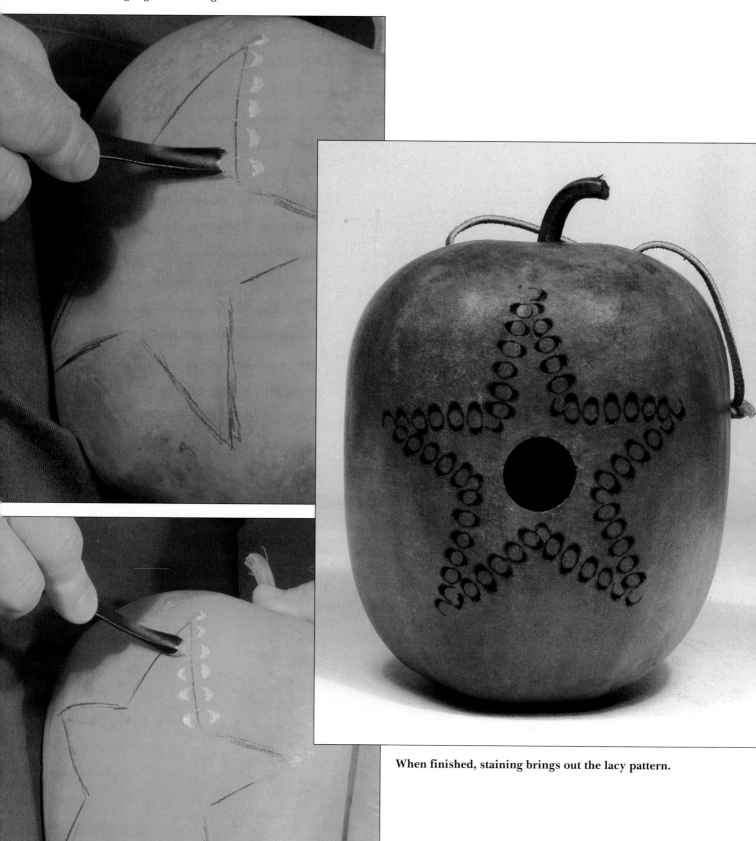

Carve the inside row.

When finished, staining brings out the lacy pattern.

Carving a Maranka Gourd

Perhaps the strangest of all gourd shapes is the maranka, also known as cave man's club, dinosaur, or dolphin (named for the dolphin-head fish, not the mammal). These gourds range in size from 8 to about 24 inches long if the handle is straight from growing on a trellis. The fascinating array of grooves and ridges are the perfect project for incorporating all your carving skills.

The maranka for this project is about 18 inches long, measuring from stem to blossom end around the curve. A close look will show that although the spaces between the ridges seem random at first glance, you can see a series of triangles, rectangles, and circles — shapes that you already have experience carving. Fill in the flat round spaces between ridges with circular carving, use band carving for rectangular areas and triangles, and add a chain of very short chips for narrow areas. In very small spaces, try adding two chips opposite each other like you made in the star pattern on the apple gourd. Or you can make the five stop cuts that touch in one point like the inside of a circle. The un-carved ridges will snake through the carving, making a nice contrast.

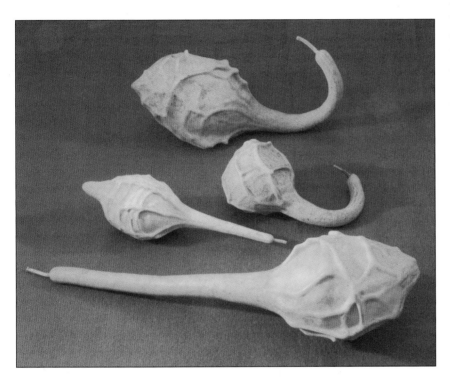

Marankas of different sizes and shapes

Carved maranka project

Find shapes to carve,
and then ...

Mark the handle.

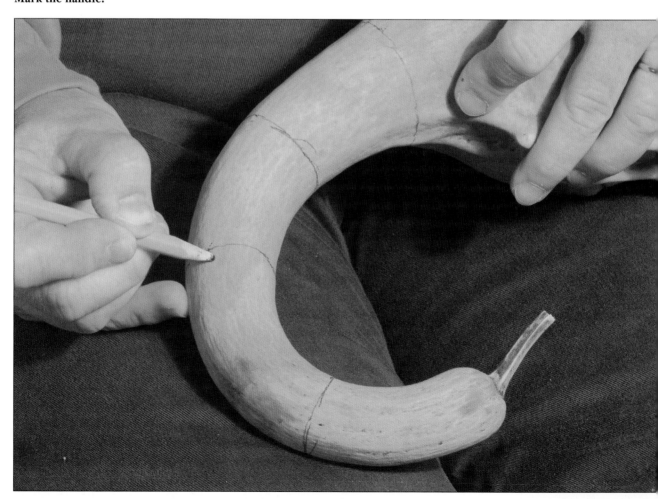

Carving the handle is not necessary — but it will be a challenge! Depending on the curvature, add three or four pencil guidelines as starting places to keep the carving looking uniform on the curves, since there will be more rows of chips on the outer curve than on the inner curve. Proceed as you did on the neck of the Indonesian bottle gourd. Again, don't stress over this since the casual observer will see the overall pattern and not where you have melded the rows. Carve carefully since the diameter of the maranka handle is small, and you have a greater chance of having the gouge slip off the handle. Leave this gourd whole — people will want to turn it over and over to see all the details!

Carved shapes within the ridges.

The finished maranka

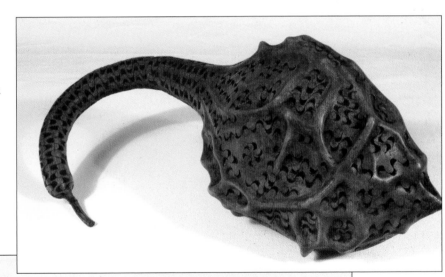

Finished maranka, second view

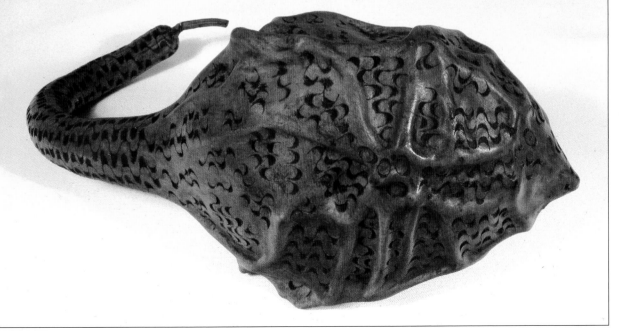

Chapter Six:
Gallery

I am happy to be able to showcase gourds carved by my teacher as well as my students. The first two pieces — the lidded bowl and the beautifully shaped Corsican flat gourd — are my favorites of the simpler pieces that my father carved. He liked to leave some portion of the gourd un-carved for contrast.

Many of my former students have become good friends and have graciously sent photographs illustrating how they transform that basic chip with their artistic skills. Enjoy their use of color and combination of carving with other media ...

Bowl. *Carved by Leslie Miller, from the collection of Joanne Stichweh.*

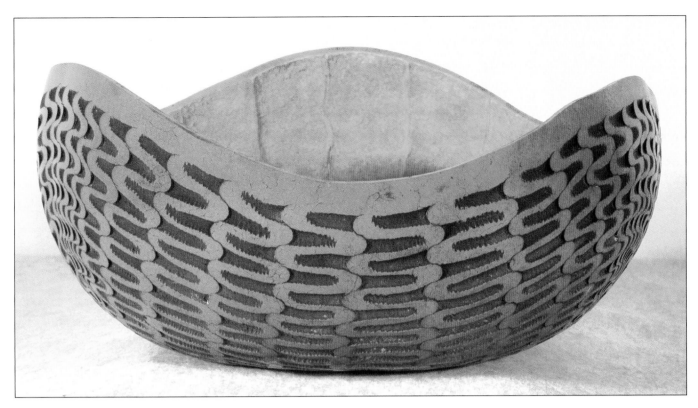

Same bowl from previous page, lid removed.

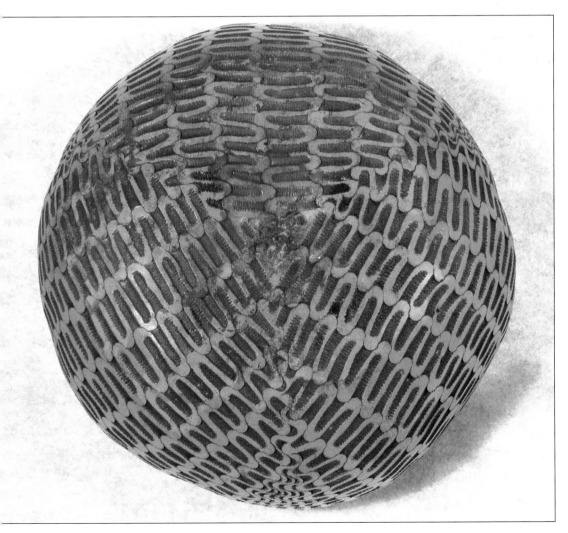

Same bowl, bottom view.

Corsican flat gourd. *Carved by Leslie Miller, from the collection of Joanne Stichweh.*

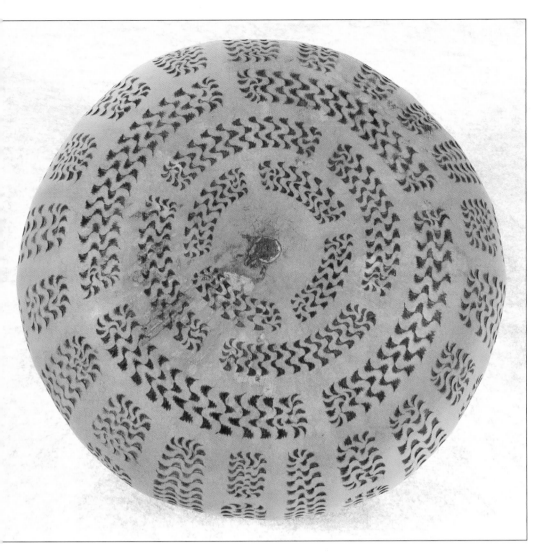

Same gourd from previous page, bottom view.

Carved and photographed by Gloria Riegel.

Carved and photographed by Gloria Riegel.

Carved and photographed by Gloria Riegel.

Carved by Rhoda Forbes and photographed by John Stacy. This is from Stacy's collection.

62

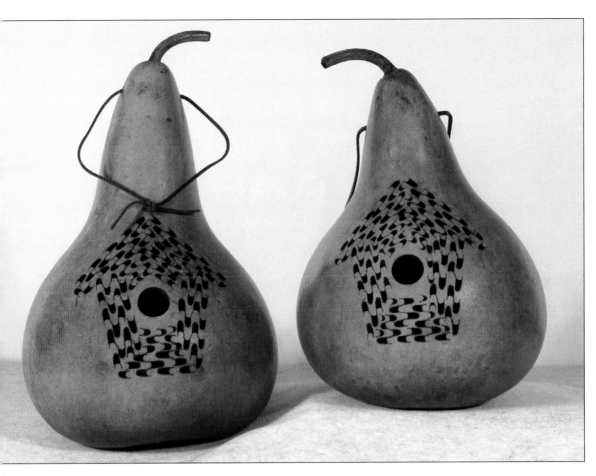

"Birdhouses."
Carved by Katie McCormick.

Ornaments. *Carved by Jeremy Rehm.*

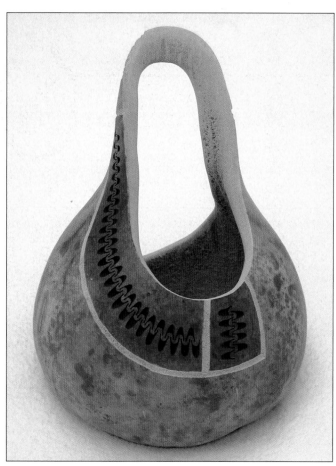

Basket. *Carved and photographed by barb cesal.*

Bowl. *Carved and photographed by barb cesal.*

"Native American Influence." *Carved and photographed by Bronii Williams.*

"Jewels of the Outback." *Carved and photographed by Bronii Williams.*

Spiral design jug. *Carved and photographed by A. B. Amis.*